Science in the News:
Critical Thinking Worksheets

Biology Connections

HOLT, RINEHART AND WINSTON

A Harcourt Classroom Education Company

Austin • New York • Orlando • Atlanta • San Francisco • Boston • Dallas • Toronto • London

To the Teacher

The video segments in the *CNN Presents Science in the News: Biology Connections* program bring the knowledge and experience of the CNN news team right into your classroom. Show students the relevance of biology to everyday life with interesting news segments that relate to the topics they are studying in class. The segments may be used as warm-up activities to start the class, as diversion points to stimulate class discussion, or as springboards for further research. Your students will see the efforts of biologists, environmentalists, students, governments, and many others who are using science to answer important questions.

These Critical Thinking Worksheets contain thought-provoking questions that require students to carefully examine the information presented in each segment. Students can sharpen their listening and critical thinking skills as they evaluate each segment. The worksheets also help to focus students' attention on the topics in each segment. You may wish to have students use their textbook as a resource for answering questions. Also, topics covered in the segment can be used as the basis for student-led discussions. The Teacher's Guide that accompanies the *Biology Connections* video contains discussion points, research ideas, and other information that will help you and your students get the most from these worksheets.

Printed in the United States of America

ISBN 0-03-056558-8

12 023 05 04

Contents

Page
Number

Segment Title

Science in the News: **Critical Thinking Worksheets**

Segment 1

Frog Pollution

1. What evidence suggested to the Oregon State researchers that ultraviolet light might be causing amphibian declines?

____

____

____

____

2. According to the scientists, why might frog eggs be exposed to increasing amounts of ultraviolet light in nature?

____

____

____

____

3. According to Michael Soule, why do scientists usually only examine one variable at a time? Why did he and his students decide to look at several variables?

____

____

____

____

4. According to the reporter, why are amphibians like the canaries that miners used in the coal mines?

____

____

____

____

Science in the News: **Critical Thinking Worksheets**

Segment 2

Cystic Fibrosis

1. The report shows a patient with cystic fibrosis wearing a "vibrating vest" that shakes her torso. What do you think is the purpose of this treatment?

____

____

____

____

2. What is the goal of gene therapy for cystic fibrosis?

____

____

____

____

3. What role did a cold virus play in the gene therapy trial described in the report?

____

____

____

____

4. Were the patients cured by gene therapy? What did the scientists learn from this preliminary experiment?

____

____

____

____

Science in the News: **Critical Thinking Worksheets**

Segment 3

Split-Liver Donor

1. Explain why doctors have had to resort to the drastic measure of splitting a donated liver in two.

____

____

____

____

2. What change in surgical technique has improved the odds of success for this procedure?

____

____

____

____

3. Do you think the same technique might work for other organs that are regularly trans-planted, such as hearts and kidneys? Explain your answer.

____

____

____

____

4. Why do you think the liver is kept on ice in a cooler while being transported to the hospital?

____

____

____

____

Science in the News: Critical Thinking Worksheets

Segment 4

Organ Cloning

1. Why does the production of headless tadpoles cause concern, according to the reporter?

____

____

____

____

____

2. Early in the segment, the reporter asks if a headless tadpole is a "Frankenstein." What does this statement mean? Why do you think the reporter chose this comparison?

____

____

____

____

3. After watching the whole segment, do you think ethicists are worried about the possibility that the tadpole experiment might be repeated on humans? Explain your answer.

____

____

____

____

____

Science in the News: Critical Thinking Worksheets

Segment 5

Cloning Mice

1. Suppose that the nucleus from the cell of a brown mouse is transferred into a white mouse's egg which has had the nucleus removed. The egg is then implanted into a black mouse. What color will the fur of the baby mouse be? Explain your answer.

2. A scientist compares the mitochondrial DNA of the newborn mouse to that of the other three mice. Which mouse should have the same mitochondrial DNA as the baby mouse? Explain your answer.

3. What benefits might cloning provide to agriculture and animal breeding?

Science in the News: **Critical Thinking Worksheets**

Segment 6

Teen Discovery

1. What are the functions of the three genes the students discovered?

____

____

____

____

2. Why would flaws in this kind of gene be likely to lead to cancer?

____

____

____

____

3. According to the first student interviewed, why is the discovery significant?

____

____

____

4. In the report, a scientist from Lawrence Livermore Laboratory says that he was eager to have help from the students. Explain why.

____

____

____

Science in the News: **Critical Thinking Worksheets**

Segment 7

Gene Progress

1. What is manic depression? What is its connection to the Human Genome Project?

____

____

____

____

____

2. How might the discovery of a genetic cause for manic depression help patients like Kay Redfield Jamison, the woman featured in the segment?

____

____

____

____

____

3. According to Francis Collins, the second scientist interviewed, why are scientists excited about the Human Genome Project?

____

____

____

____

____

4. Suppose an inherited disease runs in your family and scientists create a test for this disease. Would you want to take the test and find out whether you might develop the disease? Explain your answer.

____

____

____

____

Science in the News: **Critical Thinking Worksheets**

Segment 8

Alzheimer's Mutation

1. What is the function of the A2M gene discovered by the researchers?

____

____

____

____

2. Suppose that someone finds out that they carry the mutated version of A2M. Will they necessarily develop Alzheimer's disease? Explain your answer.

____

____

____

____

3. The reporter states that Alzheimer's disease has a special relevance for women. Explain why.

____

____

____

____

4. Quoting researchers, the reporter estimates that the creation of a drug based on this discovery might be possible in the next 10 years. Why do you think it might take this long to develop such a drug? What obstacles might scientists face during the process?

____

____

____

____

Science in the News: Critical Thinking Worksheets

Segment 9

Dino Egg Discovery

1. Many dinosaur eggs have been discovered before. What made this find unique?

____

____

____

____

2. What allowed scientists to see the embryo within the egg?

____

____

____

____

3. To be so well preserved, the eggs must have been fossilized before decaying. How might this have happened?

____

____

____

____

Science in the News: **Critical Thinking Worksheets**

Segment 10

Galápagos

1. The environment in the Galápagos differs from one island to the next. How did this variation influence the evolution of the islands' residents?

2. Like Darwin, the report notes the amazing tameness of Galápagos animals. Suggest a hypothesis that explains why extraordinary tameness evolved among these animals.

3. When the Galápagos emerged from the sea, they were lifeless lumps of lava. Now they are rich in life. Suggest several ways that animal and plant colonists might have reached the new islands from the mainland.

Science in the News: **Critical Thinking Worksheets**

Segment 11

Something Worth Saving

1. What important discoveries might biologists make in discovering and categorizing plant species in rain forests?

2. Which parts of plants can be collected by biologists to differentiate species and help classify the plants?

3. Will the efforts of biologists to identify plant species stop the clearing of rain forests? What could be done to prevent deforestation?

Science in the News: Critical Thinking Worksheets

Segment 12

Britain's Living Plan

1. List two variables considered in the World Wildlife Fund's Living Planet Index.

____

____

____

2. Describe two ways in which cars have helped drive down the index.

____

____

____

3. According to the report, the developing and developed countries share the blame for the Earth's environmental problems. How do the environmental impacts of the two kinds of countries differ?

____

____

____

4. Think about your own life. Name three ways in which your activities affect the index.

____

____

____

____

Science in the News: Critical Thinking Worksheets

Segment 13

Fat Wolves

1. What happened to the original wolf population of Yellowstone?

____

____

____

____

2. Why do you think the wolves are being kept in an "adjustment pen" before their release into the wild?

____

____

____

____

3. A rancher interviewed in the piece says he is concerned about the wolf population, even though no livestock have been killed yet. Explain why.

____

____

____

____

4. What happens if a wolf kills one of the rancher's cows?

____

____

____

____

Science in the News: Critical Thinking Worksheets

Segment 14

Tropical Reforestation

1. What are the two main reasons Costa Rica's forests have been cut down?

____

____

____

____

2. How do researchers determine which trees are best suited for reforestation efforts?

____

____

____

____

3. Identify some of the benefits of reforestation projects and of planting trees in general.

____

____

____

____

4. Will the effort described in the video segment return the forest to its prelogging condition? Explain.

____

____

____

____

   Biology Connections

Science in the News: Critical Thinking Worksheets

Segment 15

Greening Sudbury

1. What are the environmental and health effects of exposure to acidic pollution?

____

____

____

2. How did the emissions from just one industry result in such devastating pollution?

____

____

____

3. Why did the mining company build a taller smokestack before implementing more-effective pollution control measures?

____

____

____

4. How did the smelter benefit economically from the new pollution control equipment?

____

____

____

5. How has the Sudbury community helped clean up the local environment?

____

____

____

Science in the News: Critical Thinking Worksheets

Segment 16

Depression Virus

1. According to the report, how is depression different from merely "the blues"?

____

____

____

2. What are the symptoms of Borna virus infection in horses?

____

____

____

3. What two lines of evidence described in the report support the hypothesis that Borna virus causes at least some cases of depression?

____

____

____

4. What if a larger study found that Borna virus was widespread in the population? Would that weaken or strengthen the case that the virus is involved in depression? Explain your answer.

____

____

____

Science in the News: Critical Thinking Worksheets

Segment 17

Killer Algae

1. How does *Pfiesteria* affect fish? Does it affect other animals in the area?

____

____

____

____

2. Why does JoAnn Burkholder take such elaborate precautions before working with *Pfiesteria* in the lab?

____

____

____

____

3. According to Burkholder's research, what conditions favor outbreaks of *Pfiesteria* in the Neuse River?

____

____

____

____

4. How might humans be causing population explosions of *Pfiesteria*?

____

____

____

____

Science in the News: **Critical Thinking Worksheets**

Segment 18

Lethal Mushrooms

1. Why are death cap mushrooms hard to identify?

____

____

____

____

2. How does the toxin of the death cap mushroom cause death?

____

____

____

____

3. Considering the large numbers of mushrooms eaten every year, why are there not more poisonings in the United States?

____

____

____

____

4. Suppose that a mushroom enthusiast eats only those mushrooms he picked in the fall. Would this strategy guarantee his safety from poisoning? Explain your answer.

____

____

____

____

____

Science in the News: Critical Thinking Worksheets

Segment 19

Magnolia DNA

1. Explain why PCR is such a powerful tool for scientists working with DNA.

____

____

____

____

2. What must happen for DNA to survive in a fossil that is millions of years old?

____

____

____

____

3. How would having samples of ancient DNA improve our understanding of evolution?

____

____

____

____

____

Science in the News: **Critical Thinking Worksheets**

Segment 20

Migrant Bees

1. Why are farmers willing to pay for the beekeeper's service?

____

____

____

____

____

2. Why is it difficult for wild populations of bees to pollinate the crops?

____

____

____

____

____

3. According to the master beekeeper, why don't the bees simply fly away when they are left beside a field?

____

____

____

____

____

Science in the News: Critical Thinking Worksheets

Segment 21

Maple Syrup

1. Does the sap come from the xylem or the phloem? Explain your answer.

____

____

____

____

2. Why do the sap-harvesting teams use horses for transportation within the forest?

____

____

____

____

3. What happens if the sap is not boiled at the correct temperature and for the right amount of time?

____

____

____

____

4. How could the production of maple syrup help conserve forests?

____

____

____

____

____

Science in the News: **Critical Thinking Worksheets**

Segment 22

Plastic Farming

1. How did the scientist "reprogram" the plants to start making plastic?

____

____

____

____

2. At the moment, what is the biggest economic obstacle to producing plastic in this way?

____

____

____

____

3. What are some of the products that could be made from plant plastic?

____

____

____

____

4. What are some of the environmental advantages of making plastic with plants instead of with petroleum?

____

____

____

____

____

Science in the News: **Critical Thinking Worksheets**

Segment 23

Mexico Sea Turtle

1. What caused the Kemp's ridley turtle population to decline?

____

____

____

____

2. How does hatching the turtles' eggs in captivity help boost the turtle population?

____

____

____

____

3. Most Kemp's ridleys nest on a single beach in Mexico. How does this make the species more vulnerable?

____

____

____

____

Science in the News: Critical Thinking Worksheets

Segment 24

Year of the Reef

1. What percentage of the coral reefs surveyed in Reef Check showed evidence of human impact? What are some examples of that impact?

____

____

____

____

2. What were two of the unexpected findings from the reef survey?

____

____

____

3. It seems strange to hear a coral reef described as "sick." Explain how pathogens could affect a reef.

____

____

____

____

4. According to the last scientist interviewed in the piece, what actions might be needed to protect reefs?

____

____

____

____

Science in the News: **Critical Thinking Worksheets**

Segment 25

Asthma Roaches

1. According to the report, what puzzle did the scientists seek to solve with this study?

____

____

____

____

2. What did they conclude from the study?

____

____

____

____

3. What are some alternative reasons for the high rates of asthma among kids living in urban areas?

____

____

____

____

4. According to the scientist interviewed in the segment, why is discovering a link between asthma and roaches important?

____

____

____

____

Science in the News: **Critical Thinking Worksheets**

Segment 26

For the Birds

1. What eliminated the murre population, once 3,000 strong, from Devil's Slide?

2. Where does the money for the restoration effort come from?

3. Why is it necessary to go to such lengths to encourage the birds to nest on the rock?

4. How might the birds' preference for crowded nesting sites have been advantageous under normal circumstances?

Science in the News: Critical Thinking Worksheets

Segment 27

Fish Farming

1. What characteristics would make a fish a good candidate for aquaculture?

____

____

____

____

2. What are some other species besides those mentioned in the video that you think could be raised successfully using aquaculture? Why? Which species would not be good candidates?

____

____

____

____

3. Compare the food safety of eating fish caught in natural water bodies versus fish raised in aquaculture.

____

____

____

____

4. What are some of the perceived benefits and drawbacks of aquaculture?

____

____

____

____

Science in the News: **Critical Thinking Worksheets**

Segment 28

What's Slithering in Guam?

1. How did the brown tree snake arrive in Guam?

____

____

____

____

2. Why do officials not simply introduce a predator species to keep the nonnative snake species in check?

____

____

____

____

3. Why is it important to save the last breeding pair of a species?

____

____

____

____

4. What are some of the reasons that many species around the world need captive-breeding programs to ensure their survival?

____

____

____

____

Science in the News: Critical Thinking Worksheets

Segment 29

Animal Communication

1. What are three possible signals a kangaroo rat can make with its feet?

____

____

____

____

2. Suggest a hypothesis to explain how a killer whale pod might benefit from having a unique "dialect" that other pods would not understand.

____

____

____

____

3. Why do you think natural selection might favor having a different alarm call for each kind of predator?

____

____

____

____

4. What are some similarities between the animal "languages" portrayed in the segment and human languages?

____

____

____

____

Science in the News: Critical Thinking Worksheets

Segment 30

Tanning Effects

1. List three precautions you can take to reduce your risk of skin cancer.

2. Even though tanning salons use the less-damaging Ultraviolet A, why does the dermatologist in the report recommend against this method of tanning?

3. According to the researcher from the University of California, what is another motivation for reducing sun exposure?

Science in the News: **Critical Thinking Worksheets**

Segment 31

High Blood Pressure

1. High blood pressure causes or contributes to several diseases. List three.

____

____

____

____

2. What are some possible reasons so few people with high blood pressure are getting adequate care?

____

____

____

____

3. Why do you think the report recommends that people lose weight if they have high blood pressure?

____

____

____

____

4. Consider the recommendations for people with high blood pressure. Which recommendation do you think is the hardest to maintain? Which is the easiest? Explain your answers.

____

____

____

____

Science in the News: **Critical Thinking Worksheets**

Segment 32

Going Vegetarian

1. According to the first expert interviewed in the piece, why are many Americans shifting to a vegetarian diet?

____

____

____

____

2. What are some of the benefits of a diet that includes plenty of fruits and vegetables?

____

____

____

____

3. What are the nutritional pitfalls of a vegetarian diet? How can vegetarians avoid these problems?

____

____

____

____

4. The report mentions the environmental benefits of vegetarianism. Explain what these benefits are.

____

____

____

____

Science in the News: Critical Thinking Worksheets

Segment 33

Salmonella Outbreak

1. What are the main sources of salmonellosis?

____

____

____

____

2. A friend tells you that she thinks it is impossible for her to get food poisoning caused by *Salmonella* because she always cooks meat thoroughly before eating it. What would you say to her about her risks of infection?

____

____

____

____

3. What are the symptoms of food poisoning caused by a *Salmonella* infection?

____

____

____

____

4. Why do you think it is important for someone who is sick with salmonellosis to get plenty of fluids?

____

____

____

____

Science in the News: **Critical Thinking Worksheets**

Segment 34

Stroke Brain Repair

1. Why do you think scientists conducted the first studies of neuron transplantation on so few patients?

____

____

____

____

2. Why did the doctors think it was necessary to transplant neurons after a stroke?

____

____

____

____

3. What must happen before the transplanted cells can restore a function lost because of a stroke?

____

____

____

____

4. What did the researchers from the University of California discover about brain cells? Why were they so cautious about applying their findings to people?

____

____

____

____

Science in the News: Critical Thinking Worksheets

Segment 35

Obesity Hormone

1. Where in the brain is orexin produced? What does that part of the brain accomplish?

____

____

____

____

2. What evidence suggests that orexin helps control weight in rats?

____

____

____

____

3. Why do you think it will take at least 10 years to develop a drug to block orexin production?

____

____

____

____

____

Segment 1

Frog Pollution

1. Amphibian eggs died when raised in ponds that were exposed to sunlight. Eggs raised in protective containers in the lab survived.
2. The scientists theorized that the thinning of the ozone layer could be allowing more ultraviolet light to reach the pond's surface.
3. Scientists usually concentrate on a single variable because the experiments are simpler and easier to control. Soule and his students decided to look at the interactions among multiple factors because such a study better reflects what happens in nature.
4. Amphibians are often the first creatures to respond to a change in their surroundings. Like miners' canaries, they can alert us to environmental deterioration.

Segment 2

Cystic Fibrosis

1. The lungs of CF patients become clogged with thick, sticky mucus. The vest shakes this mucus loose and prevents the formation of blockages.
2. The goal is to replace the patient's defective genes with functioning ones—ultimately allowing patients to give up their restrictive daily therapy and testing.
3. Rendered harmless, the cold virus carried a normal form of the CF gene into the cells of the lungs.
4. The patients were not cured, but scientists learned that the genes could enter cells and start working.

Segment 3

Split-Liver Donor

1. There are so few donated organs that only about half of needy recipients get a liver.
2. Doctors now split the liver within the body of the donor, which reduces the damage to the liver during removal.
3. No, the technique would not work. A divided heart or kidney could not function.
4. Since the organ has just been removed from a donor, it could start to decompose. Cooling the liver slows its deterioration.

Segment 4

Organ Cloning

1. He says the technology could be applied to humans to create "bundles of organs for harvest."
2. The reporter is asking, in essence, if the tadpole experiment has produced a monster like the Frankenstein of the movies, who was stitched together from body parts of the dead. Frankenstein symbolizes science that has gone beyond the standards of society, and the reporter is asking if this experiment has also exceeded acceptable standards.
3. No, ethicists are not worried because a human application is too far removed from what is currently possible.

Segment 5

Cloning Mice

1. The mouse's fur will be brown because the clone is a carbon copy of the nuclear donor.
2. The mitochondrial DNA of the offspring comes from the egg and is independent of the nucleus. So, in this case, the white mouse will have the matching mitochondrial DNA.
3. Animals that have desirable traits can be duplicated through cloning. There is a better chance that the offspring will have the trait than with crossbreeding.

Segment 6

Teen Discovery

1. The genes help regulate cell division.
2. Cancer is the out-of-control multiplication of cells. Cell division is normally tightly regulated by genes. Releasing the genetic controls on cell division is a prerequisite for cancer.
3. Knowing how the genes go wrong might make it possible to fix the error that leads to cancer.
4. There are about 70,000 genes, most of whose functions remain unknown. The more participants analyzing genes the better, according to the scientist.

Segment 7

Gene Progress

1. Manic depression is a mental illness characterized by wild mood swings, from feelings of ecstasy to feelings of despair. Scientists involved in the project have discovered a possible genetic link to the disease, a gene or cluster of genes on Chromosome 18.
2. Knowing what gene causes the disorder could improve diagnosis, yielding more accurate tests, or it could allow scientists to design better treatments based on their knowledge of what the gene does.
3. The public is excited because of the project's potential to cure many inherited diseases.
4. Answers will vary but should be logical. Many people would want to know what they are up against, genetically speaking, but others would prefer not to know.

Segment 8

Alzheimer's Mutation

1. According to the reporter, the gene helps keep synapses functioning.
2. No, having the gene indicates you are at greater risk, but does not doom you to the disease.
3. Women constitute the majority of the over-82 age group that is particularly susceptible to the disease. Women are also more often caregivers for Alzheimer's patients.
4. Students might mention several of the following reasons: The discovery first must be replicated in other Alzheimer families. Then scientists would have to find a compound with the right activity that was not toxic and that could get to the correct part of the brain. They would then have to put the drug through safety tests and gain government approval.

Segment 9

Dino Egg Discovery

1. The embryo, on the verge of hatching, was so well preserved that many of its features were visible in the CT scan. According to one paleontologist interviewed, dinosaur embryos usually turn to mush before they fossilized.
2. The scientists used an improved CT scanner with higher resolution.
3. If the eggs were quickly buried in a layer of sediment, perhaps by a flash flood, the conditions would be right for fossilization.

Segment 10

Galápagos

1. It spawned the great diversity of species that so impressed Darwin. Populations on different islands adapted to local conditions and diverged from each other, giving rise to many species.
2. Answers will vary. Students might hypothesize that because the Galápagos have few predators, the animals no longer needed to waste energy on being constantly alert.
3. For some animals, the trip was not so hard: birds could fly, and seals could swim. Other animals, like tortoises and lizards, might float on clumps of debris. Plant seeds could float or be carried by birds.

Segment 11

Something Worth Saving

1. In addition to potential medicinal properties, tropical species may have components that could be used in insecticides.
2. Leaves, flowers, fruits, and seeds are commonly used by biologists to help classify plants.
3. No, unless governments pass legislation setting aside protected areas, the over-population of humans will cause the continued destruction of valuable rain forests. Deforestation could be prevented through population control and government regulations prohibiting the clearing of rain forests.

Segment 12

Britain's Living Plan

1. Several answers are possible, including population, freshwater supplies, pollution, forested areas, and the health of wild populations.
2. Several answers are possible. Cars create pollution and contribute to global warming. Making the steel, plastic, glass, and other components of the car consumes resources and generates pollution. Running a car takes gasoline, made from crude oil in a pollution-creating process. Obtaining crude oil causes environmental damage. Cars need roads, which eat up land.
3. Developed countries have small populations but a large appetite for resources, and they produce large amounts of pollution. In the developing countries, each person has a much smaller environmental impact, but it is summed over a much larger population.
4. Answers will vary. Students may point out ways that they use resources or create pollution or ways that they save resources, such as cycling or walking to school.

Segment 13

Fat Wolves

1. Every wolf was killed by government bounty hunters as part of a wolf eradication program intended to protect livestock.
2. This gives the wolves time to get accustomed to their surroundings—including the other wolves they will be living with. It also allows the scientists to keep an eye on the wolves during the transition to a new habitat.
3. He fears that once wolves have reduced the populations of their wild food—elk and buffalo—they will hunt easy-to-catch livestock.
4. A private fund will compensate him for the loss.

Segment 14

Tropical Reforestation

1. The forests have been cut down for timber and to create pastures for raising cattle.
2. Researchers determine how quickly each species grows in different soil types. They also determine which wood could be sold for profit.
3. Reforestation takes the pressure off the remaining forests, which might be logged otherwise. It also creates habitats for animals and other plants. Also, forests draw carbon dioxide out of the atmosphere, help regulate rainfall, protect the soil, and produce oxygen.
4. The effort will probably not return the forest to its original condition. Trees will be selected for swift growth, not because they were present in the original forest. The other plants and animals that lived in the forest before it was cut down might not return.

Segment 15

Greening Sudbury

1. For the environment, acid lowers the pH of the soil, killing plants, and it can kill aquatic life as well. People living near acid-emitting industrial plants have a higher risk of respiratory diseases.
2. Smelting to release metal ore spewed sulfur dioxide into the air. Sulfur dioxide combined with water in the atmosphere to create acid rain, which fell back to earth and poisoned the land and water. Sulfur dioxide escaping from the smelter was also toxic.
3. Taller smokestacks spread the sulfur dioxide gas over a larger area. However, this did not solve the problem; it just dispersed it.
4. The plant now captures most of the sulfur in the ore and turns it into sulfuric acid, which can be sold.
5. The community helped spread lime to neutralize the acid in the soil. They also helped with replanting.

Segment 16

Depression Virus

1. Depression is deep, long-term despair, not just a short bout of low spirits.
2. The virus attacks the part of the brain responsible for emotions. The horses stop eating, walk in circles, and often die.
3. A strain of Borna virus has been found in the blood of depressed patients. Also, some patients treated with an antiviral drug have recovered from their depression.
4. If many people had the virus but did not have depression, that would weaken the case that Borna is involved in depression.

Segment 17

Killer Algae

1. It kills fish, eating away their flesh. It causes illness in people, but its effects on other animals, such as birds that eat the fish, are unknown.
2. The alga releases a toxin that causes a range of symptoms in humans, including skin ulcers, memory loss, confusion, nausea, bone aches, and shortness of breath. She was once exposed and lost her short-term memory.
3. The algae seem to thrive when phosphorus and nitrogen levels are high in river water.
4. Growing numbers of farms, homes, and industrial developments along the river are releasing large amounts of waste that contain nitrogen and phosphorus into the river.

Segment 18

Lethal Mushrooms

1. Their color can vary from greenish-white to gray, yellow, green and purple, which makes it hard for nonexperts to make an identification.
2. It kills by destroying the liver.
3. According to the report, wholesalers only buy from reputable, experienced pickers and reject any mushrooms that are not definitely from safe species.
4. No, the mushroom expert in the story points out that death cap mushrooms can be found all year round in some areas.

Segment 19

Magnolia DNA

1. Starting with just a few molecules of DNA, PCR can make billions of copies that researchers can sequence.
2. In essence, decay must slow down so much that the DNA is not destroyed.
3. Scientists could determine the molecular characteristics of past species, instead of inferring them based on the characteristics of modern species. Scientists also might be able to correlate structural changes with the molecular alterations that caused them.

Segment 20

Migrant Bees

1. Without pollination by bees, the crops could not yield any produce that the farmers could sell. Their livelihood depends on bees.
2. Pesticides have made them so rare that there are not enough individual bees to pollinate all the plants.
3. The beekeeper says the bees take note of the hive's position and, after working, return like homing pigeons.

Segment 21

Maple Syrup

1. Because it contains sugars, the sap must come from the phloem, the tree's sugar-transporting system.
2. The terrain is rugged, but tractors would cause soil erosion and rip up the roots of the maple trees. Horses are less destructive.
3. Its consistency will be wrong—it will either crystallize or ferment, making it useless.
4. Sap comes from live trees, so there is an incentive to let the forest stand instead of cutting it down.

Segment 22

Plastic Farming

1. He inserted genes from soil bacteria into existing biochemical pathways for storing oils and fats.
2. Although the scientist boosted the yield per acre by about 100 times, the process is still slightly more expensive than traditional plastic manufacturing techniques.
3. The scientist mentions milk jugs, and the reporter adds shampoo bottles, bandages, and sutures.
4. First, the plastic is biodegradable and non-toxic. Second, the plastic does not require crude oil, thus reducing pollution and environmental damage from oil drilling, transportation, and refining. Third, unlike oil, the plants are a renewable source of raw material for plastic.

Segment 23

Mexico Sea Turtle

1. Natural predators and human actions caused the decline. The shrimp industry was to blame for part of the decline. Though the report does not specify how, large numbers of turtles became ensnared in shrimping nets and drowned.
2. The eggs are protected from predators; thus, more young turtles make it to the sea.
3. Any disruption to this key piece of habitat can threaten the species' ability to reproduce itself.

Segment 24

Year of the Reef

1. One hundred percent of reefs showed human impact, including damage from pollution, overfishing, and boats dropping anchor.
2. One was the large number of reef diseases, like White Death and Black Band, that turned up in the Caribbean and the Florida Keys. The other was the unexpectedly high impact from fishing on reef health.
3. The reef is built by living coral animals and also depends on many other inhabitants, such as fish and shellfish, for its existence. Pathogens can attack any of these creatures and cause damage to the reef.
4. We might need to set aside reef areas and prohibit fishing and discharge of boat wastes. We also need to find out how to deal with wastes that are produced on land and that can enter the ocean and damage reefs.

Segment 25

Asthma Roaches

1. The scientists wanted to know why urban children have three times more hospital visits for asthma than their suburban peers.
2. The scientists concluded that many of the cases of asthma in urban kids are triggered by exposure to cockroaches.
3. Some alternative reasons are smog, limited health care, allergies to pets, and smoking by parents.
4. By implicating roaches as a cause of asthma, cleaning up homes to remove roaches and their allergens, should reduce the incidents of asthma.

Segment 26

For the Birds

1. Gillnetting killed many of the birds; the remainder fell victim to a series of oil spills in the 1980s.
2. A $5 million legal settlement from one of the spills.
3. Murres like nesting in dense flocks. They do not want to nest on the rock unless they see large numbers of other birds there.
4. It may be a case of safety in numbers. A bird can hide its nest among many others and gain protection from predators.

Segment 27

Fish Farming

1. Answers will vary. The ideal fish would be small enough to be raised in limited space, fast-growing, resistant to disease and parasites, easy and inexpensive to raise, tasty, and marketable.
2. Answers will vary. Students may list small, fast-growing species like bass, oysters, and crawfish as good candidates. Unsatisfactory species could include those like sharks that are too large or those that grow slowly.
3. Aquaculture reduces the risk of contamination because the fish's environment can be closely regulated.
4. Benefits include fish and shellfish that are free of disease and chemicals. Drawbacks include the high costs and difficulties of farm raising. Also, some people believe that farm-raised fish do not taste as good as fish caught in the wild.

Segment 28

What's Slithering in Guam?

1. The brown tree snake got to Guam as a stowaway in Navy machinery and bicycles.
2. The predator might also destroy native species.
3. The species can never rebound if there are no more individuals that can reproduce. The last breeding pair is the species' last chance.
4. As with the local bird species in Guam, other species worldwide are becoming extinct in the wild. Captive breeding programs are the only way to keep such species going.

Segment 29

Animal Communication

1. A kangaroo rat sends unique signals for mating, territorial encounters and to warn when predators are near.
2. Answers will vary. One possibility is that pod members can communicate without other pods eavesdropping. If a member of a pod finds food, for instance, it can alert all the other members of the group without notifying outsiders who might steal the food.
3. Specific alarm calls give the animals more information for choosing a means of escape. For instance, taking wing is a good way to get away from a snake, but it increases a bird's vulnerability to hawks.
4. Answers will vary. Students may mention that both kinds of languages have dialects with regional accents. Both have an effect on the behavior of others. And in both, the "signals" depend on the situation.

Segment 30

Tanning Effects

1. Reduce sun exposure, especially during the period of peak UV intensity between 10 A.M. and 2 P.M. Use sunscreen with an SPF of at least 15. Cover skin with hats and clothes when you are in the sun.
2. Ultraviolet A still damages the skin and can cause cancer.
3. Sunlight also causes skin to age faster and is responsible for 90 percent of unwanted blemishes like wrinkles and liver spots.

Segment 31

High Blood Pressure

1. The segment names heart attack, stroke, and kidney disease.
2. Many probably do not know about their high blood pressure, others may not be taking the steps their doctor recommended or may not be able to afford treatment, and some doctors may not be prescribing adequate treatments.
3. The heavier you are, the more tissue you have and the harder your heart must work to supply blood to that tissue. Over the long term, the heart can be weakened by this extra burden.
4. Answers will vary but should be logical. For instance, students might say it would be hardest to get regular exercise because many people are so busy or easiest to take prescribed medication.

Segment 32

Going Vegetarian

1. There are three reasons Americans are shifting to a vegetarian diet: health, environmental concerns, and religious beliefs.
2. Studies have found that people who eat this kind of diet have lower rates of hypertension, heart attack, certain cancers, and obesity.
3. A vegetarian diet may contain too much dairy or nuts, causing weight gain. And the diet may not provide enough of certain nutrients such as calcium, zinc, iron, and vitamin B-12. Vegetarians can avoid the first problem by limiting their consumption of dairy and nuts. They can compensate for the second problem by taking supplements or eating fortified foods.
4. If more people were vegetarians, pollution, overgrazing and rangeland damage from livestock could be reduced.

Segment 33

Salmonella Outbreak

1. The main sources of salmonellosis are uncooked meats, unpasteurized milk, and eggs, and food tainted by cross-contamination.
2. Cooking food thoroughly is one key to preventing infection. But the bacterium can be spread by contact with raw foods or by utensils, kitchen counters, and people's hands that have been contaminated.
3. The symptoms of salmonellosis infection are nausea, painful abdominal cramps, diarrhea, and vomiting.
4. The person is losing fluids through diarrhea and vomiting, so he or she must drink fluids to prevent dehydration.

Segment 34

Stroke Brain Repair

1. The studies were intended to determine the safety of the procedure. The small number of patients is a precaution in case the treatment has serious side effects.
2. Strokes kill neurons, and the brain was thought incapable of regenerating neurons.
3. The cells must "settle in" and link up with existing brain cells.
4. The researchers discovered that after a stroke, certain cells in rodent brains could differentiate into neurons and help reconnect existing neurons. They were cautious because the results came only from rodent studies. They do not know that this happens in people.

Segment 35

Obesity Hormone

1. Orexin is made in the hypothalamus, which regulates many other body functions.
2. Rats given injections of orexin ate 6 to 10 times as much food as normal rats.
3. The scientist in this segment has not done any work on humans, and he must still find a safe drug that blocks the action of orexin.